JUMBO COLORING BOOK

A FOREST ECOSYSTEM IS A COMMUNITY OF LIVING THINGS-LIKE PLANTS, ANIMALS, AND MICROBES-THAT DEPEND ON EACH OTHER AND THEIR ENVIRONMENT, COVERING ABOUT ONE-THIRD OF EARTH'S SURFACE.

I0529144

Biographical Note
Forest Ecosystems An Educational Coloring Book is a new work,
first published by Little Artist Studio in 2025.

International Standard Book Number
ISBN 979-8-9992504-2-1

www.littleartiststudio.org

EXPLORE THE WORLD'S MAJESTIC FORESTS THROUGH OVER 85 BEAUTIFULLY DETAILED COLORING PAGES. THIS EDUCATIONAL COLORING BOOK SHOWCASES ICONIC FOREST ECOSYSTEMS FROM AROUND THE GLOBE, DESIGNED TO INSPIRE LEARNING AND A DEEPER APPRECIATION FOR OUR PLANET'S DIVERSE ENVIRONMENTS. PART OF LITTLE ARTIST STUDIO'S ACCLAIMED EDUCATIONAL SERIES, EACH FULL-PAGE ILLUSTRATION TELLS A VIVID STORY THAT IGNITES CURIOSITY AND CREATIVITY. WITH SINGLE-SIDED PAGES, ARTISTS OF ALL AGES CAN USE ANY COLORING MEDIUM AND EASILY DISPLAY THEIR FINISHED ARTWORK. PERFECT FOR NATURE LOVERS, EDUCATORS, AND CREATIVE MINDS ALIKE.

TEMPERATE FORESTS

MAPLE TREES GROW IN COOL PLACES CALLED TEMPERATE FORESTS. THEIR LEAVES ARE WIDE AND FLAT - AND THEY FALL, AND CHANGE COLOR IN THE AUTUMN SEASON.

TEMPERATE FORESTS

TEMPERATE FORESTS

IT CAN TAKE 40 GALLONS OF SAP TO MAKE JUST 1 GALLON OF MAPLE SYRUP!

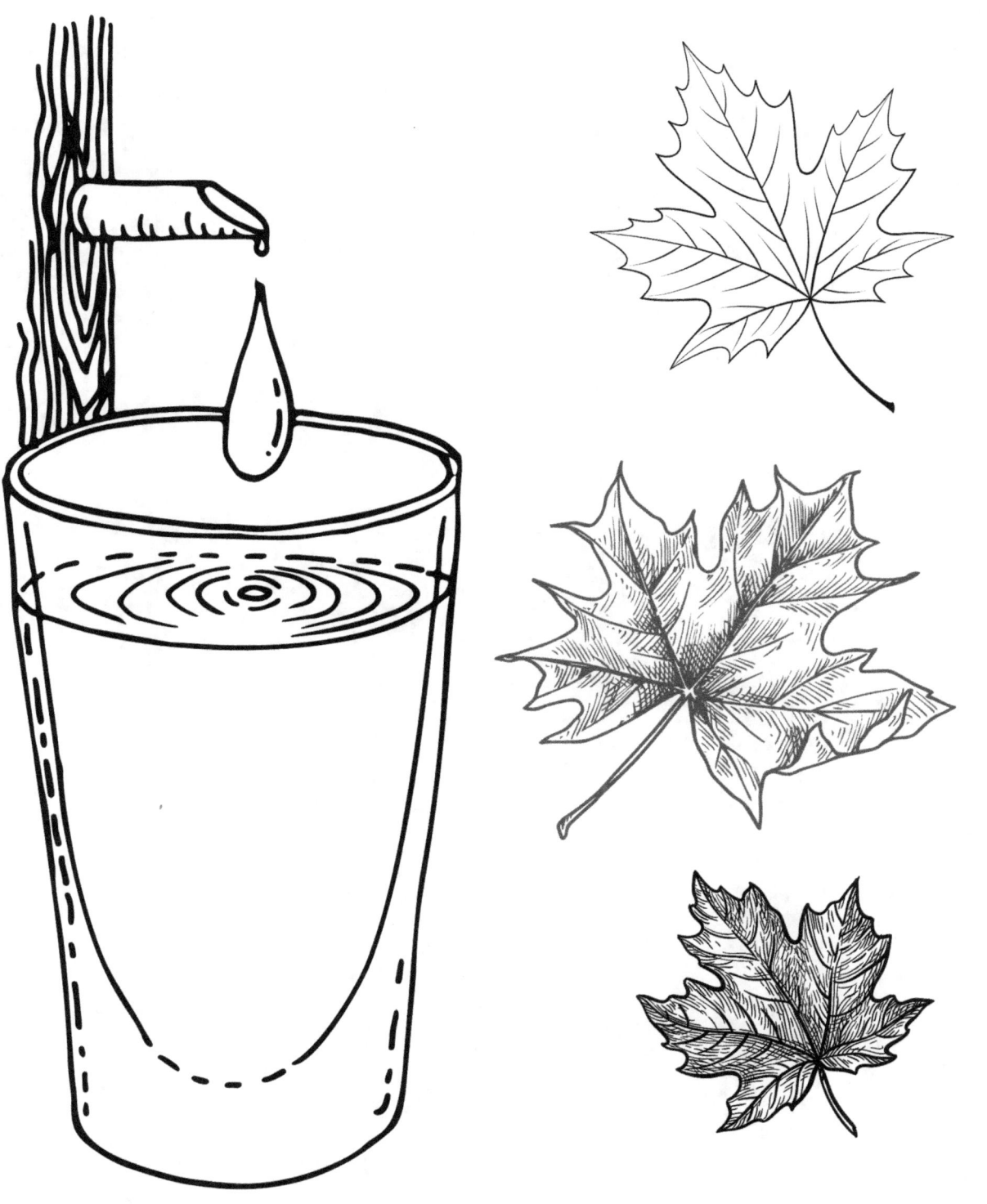

TEMPERATE FORESTS

TEMPERATE FORESTS

BIRCH TREES ARE EASY TO SPOT BECAUSE
THEY HAVE SMOOTH WHITE OR SILVER
BARK AND TALL, THIN SHAPES

TEMPERATE FORESTS

SOME TEMPERATE FORESTS HAVE
TREES WITH NEEDLES INSTEAD OF
FLAT LEAVES.

TEMPERATE FOREST

REDWOOD TREES ARE NATIVE TO TEMPERATE COASTAL FORESTS. WHILE THEY HAVE BEEN PLANTED IN SOME TROPICAL REGIONS, THEY DO NOT NATURALLY OCCUR IN TROPICAL RAINFOREST ENVIRONMENTS.

TEMPERATE FORESTS

IN THE FALL, THE LEAVES ON MANY TREES
IN TEMPERATE FORESTS CHANGE COLOR
AND GENTLY FALL TO THE GROUND.

TEMPERATE FORESTS

THESE FORESTS GET LOTS OF RAIN,
WHICH HELPS THE SOIL STAY
HEALTHY AND GROW MANY KINDS OF
TREES LIKE MAPLE, OAK, AND BIRCH.

TROPICAL FORESTS

A TROPICAL FOREST IS A WARM, RAINY FOREST FOUND NEAR THE EQUATOR.

TROPICAL FORESTS

TROPICAL FORESTS GROW IN WARM
PLACES NEAR THE EQUATOR, LIKE
SOUTHEAST ASIA, AFRICA, AND
CENTRAL AMERICA.

TROPICAL FORESTS

RUBBER TREES (HEVEA BRASILIENSIS) ARE
TROPICAL TREES KNOWN FOR PRODUCING
NATURAL RUBBER, A STRETCHY, DURABLE
MATERIAL.

TROPICAL FORESTS

RUBBER TREES PRODUCE NATURAL RUBBER, WHICH IS USED TO MAKE THINGS LIKE TIRES AND SHOES.

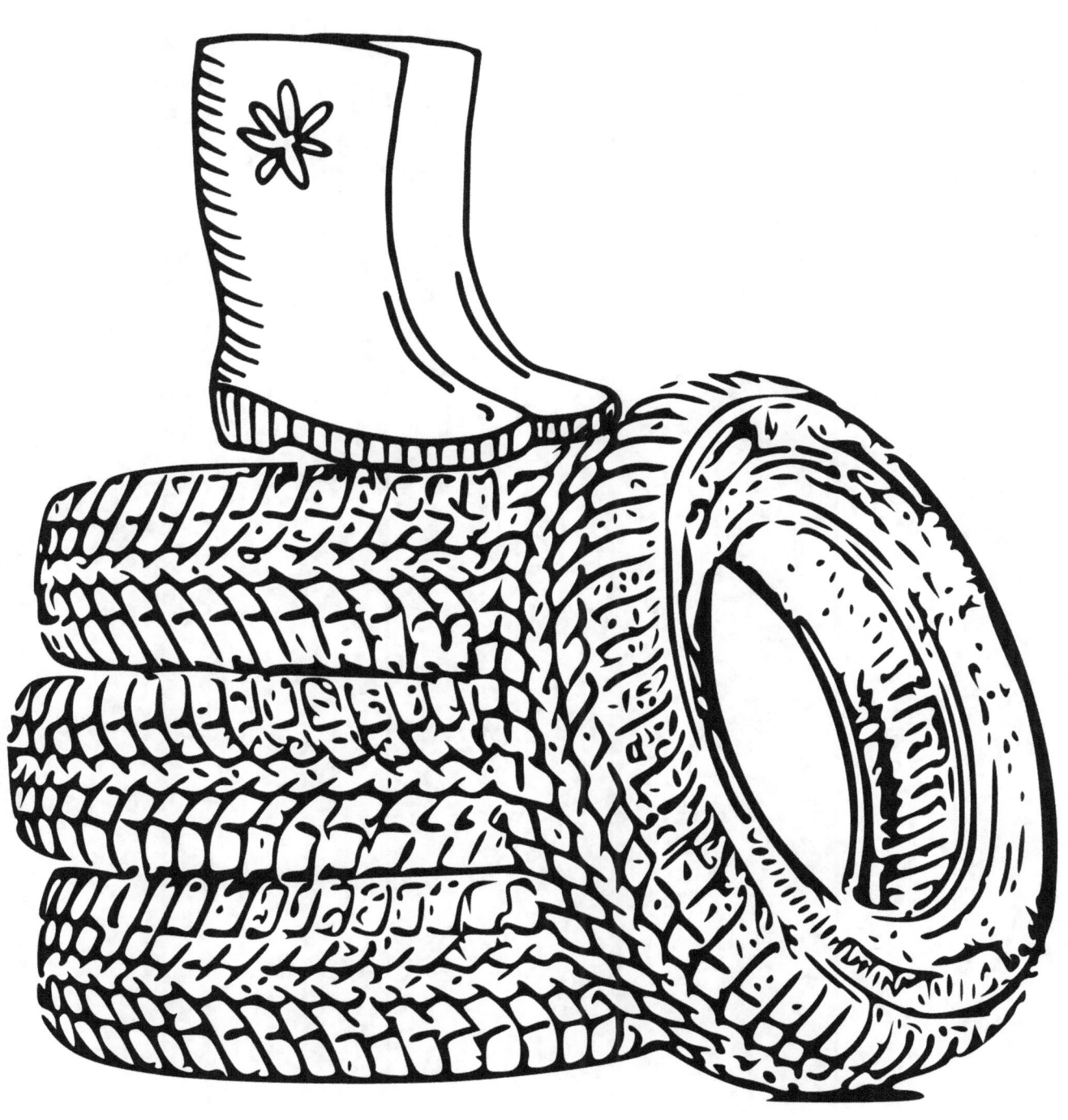

TROPICAL FORESTS

ALTHOUGH "RUBBER" DUCKS WERE ORIGINALLY
INFLUENCED BY RUBBER TECHNOLOGY AND GOT THEIR
NAME FROM EARLY VERSIONS MADE OF NATURAL
RUBBER, MOST MODERN VERSIONS ARE NOW MADE
FROM VINYL OR OTHER PLASTICS.

POTTED-RUBBER PLANT

RUBBER DUCK

TROPICAL FORESTS

THE CACAO TREE (THEOBROMA CACAO) IS AN
EVERGREEN TREE THAT PRODUCES COCOA BEANS,
WHICH ARE HARVESTED TO MAKE CHOCOLATE
AND OTHER COCOA-BASED PRODUCTS.

TROPICAL FORESTS

MANGO TREE GIVES US
DELICIOUS MANGOES AND HAS
BIG, SHINY LEAVES

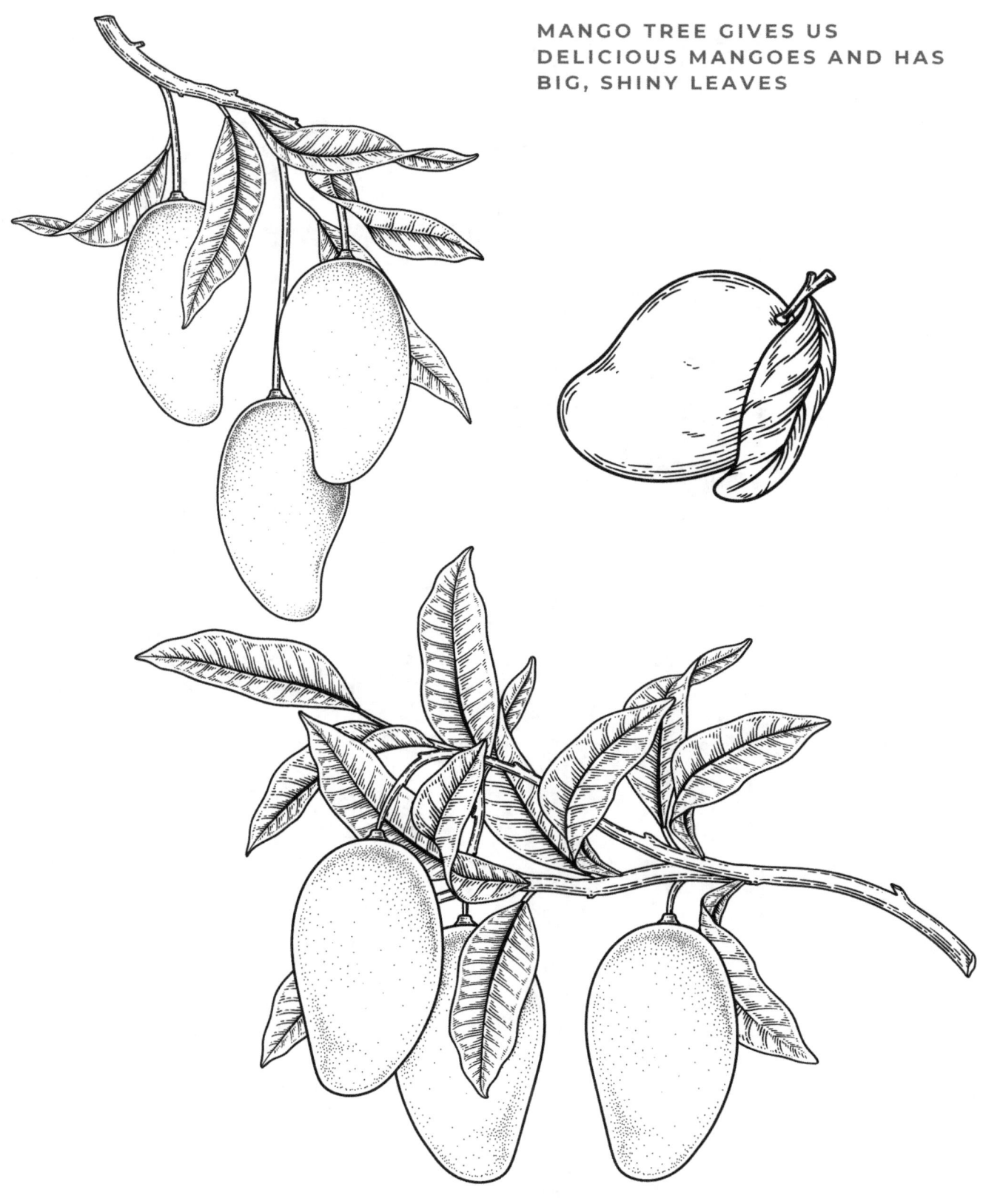

TROPICAL FORESTS

SOME ANIMALS, LIKE TOUCANS, JAGUARS, AND TREE FROGS, LIVE ONLY IN TROPICAL FORESTS AND NOWHERE ELSE IN THE WORLD.

TROPICAL FORESTS

TROPICAL FORESTS

TROPICAL FORESTS

EAGLES INHABIT TROPICAL RAINFORESTS, WITH SOME OF THE LARGEST SPECIES LIKE THE HARPY EAGLE AND PHILIPPINE EAGLE RESIDING THERE.

TROPICAL FORESTS

EAGLES HAVE SUPER SHARP EYES-THEY CAN SEE UP TO 8 TIMES BETTER THAN HUMANS! THAT MEANS THEY CAN SPOT THEIR FOOD FROM WAY UP HIGH IN THE SKY.

TROPICAL FORESTS

THESE POWERFUL BIRDS OF PREY ARE APEX
PREDATORS, HUNTING SLOTHS, MONKEYS, AND
OTHER ANIMALS WITHIN THE FOREST CANOPY

TROPICAL FORESTS

AFRICAN MAHOGANY TREES ARE TALL AND HAVE
SUBSTANTIAL TRUNKS. THEY CAN REACH
HEIGHTS UP TO 100-200 FEET, AND TRUNKS UP
TO 6 FEET IN DIAMETER.

TROPICAL FORESTS

AFRICAN MAHOGANY IS HIGHLY REGARDED FOR ITS
BEAUTY, DURABILITY, AND WORKABILITY, MAKING IT A
POPULAR CHOICE FOR HIGH-END FURNITURE AND
CABINETRY.

BOREAL FOREST

BOREAL FORESTS, ALSO CALLED TAIGA, ARE THE COLDEST FORESTS ON EARTH!

BOREAL FOREST

BOREAL FOREST

BOREAL FOREST

BEAR

FOX

BOREAL FOREST

BOREAL FORESTS ARE HOME TO ANIMALS LIKE MOOSE, WOLVES, AND SNOWSHOE HARES.

BOREAL FOREST

WOLF

BOREAL FOREST

HARE

BOREAL FOREST

DEER

BOREAL FOREST

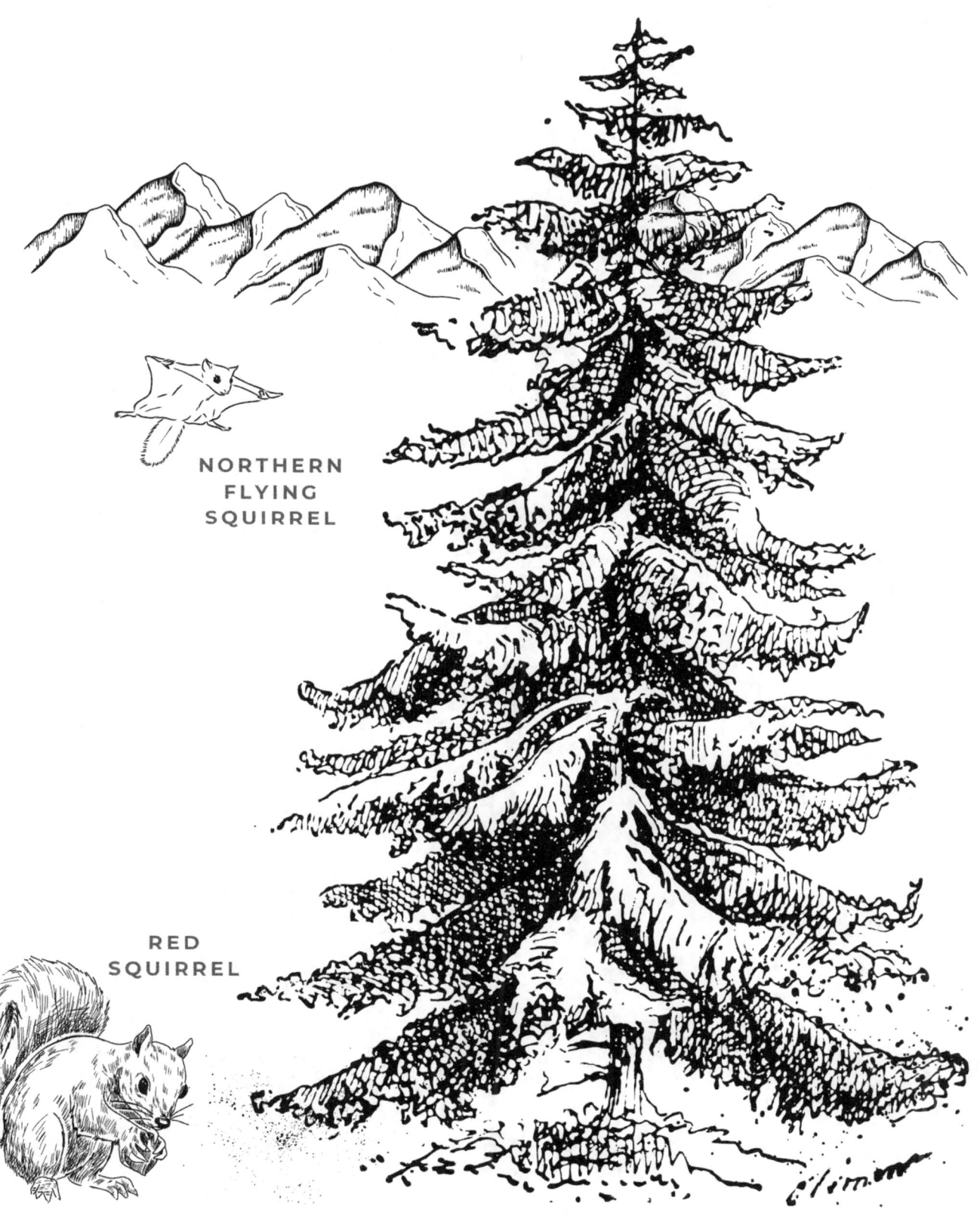

NORTHERN
FLYING
SQUIRREL

RED
SQUIRREL

BOREAL FOREST

AMAZON RAINFOREST

THE AMAZON RAINFOREST IS A
FAMOUS BIODIVERSITY HOTSPOT.

CAPUCHIN MONKEY

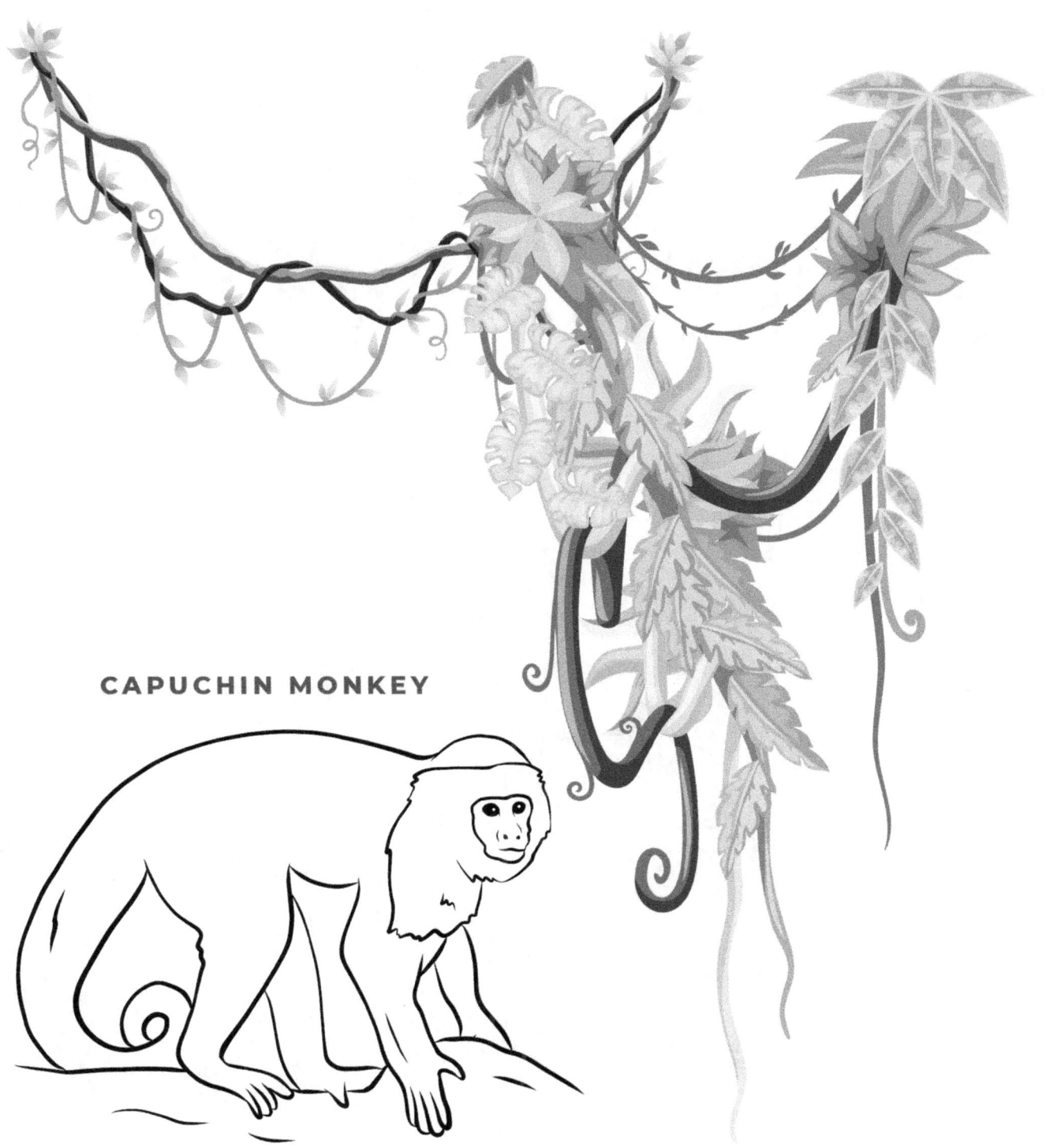

AMAZON RAINFOREST

A BIODIVERSITY HOTSPOT IS A PLACE ON EARTH THAT HAS LOTS OF DIFFERENT PLANTS AND ANIMALS, MANY OF WHICH CAN'T BE FOUND ANYWHERE ELSE.

SQUIRREL MONKEY

AMAZON RAINFOREST

THE AMAZON IS SOMETIMES CALLED THE "LUNGS OF THE PLANET" BECAUSE IT HELPS PRODUCE OXYGEN AND KEEPS THE CLIMATE STABLE.

LIZARD

IGUANA

AMAZON RAINFOREST

TOUCAN

AMAZON RAINFOREST

POISON DART FROGS ARE TINY, BUT
BRIGHTLY COLORED TO WARN OTHER
ANIMALS THAT THEY ARE POISONOUS

AMAZON RAINFOREST

THREE-TOES SLOTHS
CAN BE FOUND IN
THE RAINFORESTS
SURROUNDING
KAIETEUR FALLS IN
GUYANA, SOUTH
AMERICA.

AMAZON RAINFOREST

RIVER TURTLE

AMAZON RAINFOREST

THE AMAZON RAINFOREST IS HOME TO THE AMAZON RIVER AND ITS NUMEROUS TRIBUTARIES.

PARROT

AMAZON RAINFOREST

MAJOR RAINFOREST RIVERS INCLUDE THE AMAZON—
THE LARGEST BY VOLUME—ALSO TRIBUTARIES SUCH
AS THE MADEIRA, PURUS, JURUÁ, XINGU, AND
TAPAJÓS - ALL VITAL TO THE RAINFOREST'S
ECOSYSTEM AND BIODIVERSITY

AMAZON RAINFOREST

WHILE NOT THE HIGHEST WATERFALL OVERALL, THE KAIETEUR FALLS IS TALLER THAN NIAGARA FALLS AND BOASTS A SIGNIFICANT WATER FLOW, MAKING IT A SPECTACULAR NATURAL WONDER.

AMAZON RAINFOREST

CACHOEIRA DE IRACEMA (IRACEMA WATERFALL) IS A SCENIC WATERFALL NEAR PRESIDENTE FIGUEIREDO IN AMAZONAS, BRAZIL. IT'S A POPULAR ECO-TOURISM SPOT, KNOWN FOR ITS NATURAL BEAUTY AND EASY ACCESS.

AMAZON RAINFOREST

WITH ITS GENTLE CURRENTS AND SHALLOW POOLS, THE CACHOEIRA ASFRAMA WATERFALL IN THE AMAZON RIVER IS SUITABLE FOR FAMILIES WITH CHILDREN. IT'S ALSO A FAVORED SPOT FOR RELAXATION AND NATURE WALKS.

AMAZON RAINFOREST

THE AMAZON RAINFOREST IS HOME TO AN INCREDIBLE VARIETY OF ANIMALS—MANY OF THEM FOUND NOWHERE ELSE ON EARTH.

BLACK CAIMAN FOUND MAINLY IN THE AMAZON BASIN.

AMAZON RAINFOREST

THE GIANT MONKEY FROG (PHYLLOMEDUSA BICOLOR) IS A LARGE, GREEN AMAZONIAN TREE FROG KNOWN FOR ITS USE IN THE KAMBÔ RITUAL FOR HEALING AND PURIFICATION.

TROPICAL FOREST

THE PROBOSCIS MONKEY LIVES ONLY IN THE SWAMPY LOWLAND RAINFORESTS OF COASTAL AREAS OF BORNEO AND THE MENTAWAI ISLANDS. THESE LARGE MONKEYS ARE BEST KNOWN FOR THE MALE'S LARGE PROTRUDING NOSE.

TROPICAL FOREST

MANDRILLS ARE NATIVE TO THE TROPICAL
FORESTS OF WEST-CENTRAL AFRICA, SPECIFICALLY
IN PARTS OF CAMEROON, EQUATORIAL GUINEA,
GABON, AND THE REPUBLIC OF THE CONGO.

FOREST FLOOR

THE FOREST FLOOR IS THE BOTTOM LAYER OF THE FOREST, AND IT'S FULL OF LIFE!

FOREST FLOOR

DEAD LEAVES AND PLANTS DECOMPOSE
AND TURN INTO FOOD FOR THE SOIL,
HELPING NEW PLANTS GROW.

FOREST FLOOR

TREE ROOTS HELP HOLD THE SOIL
TOGETHER, PREVENT EROSION, AND TAKE
IN WATER AND NUTRIENTS TO KEEP
PLANTS HEALTHY.

FOREST FLOOR

FUNGI, LIKE MUSHROOMS AND MOLDS, GROW
ON DECAYING PLANT MATERIAL. THEY HELP
BREAK IT DOWN AND RECYCLE NUTRIENTS
BACK INTO THE SOIL.

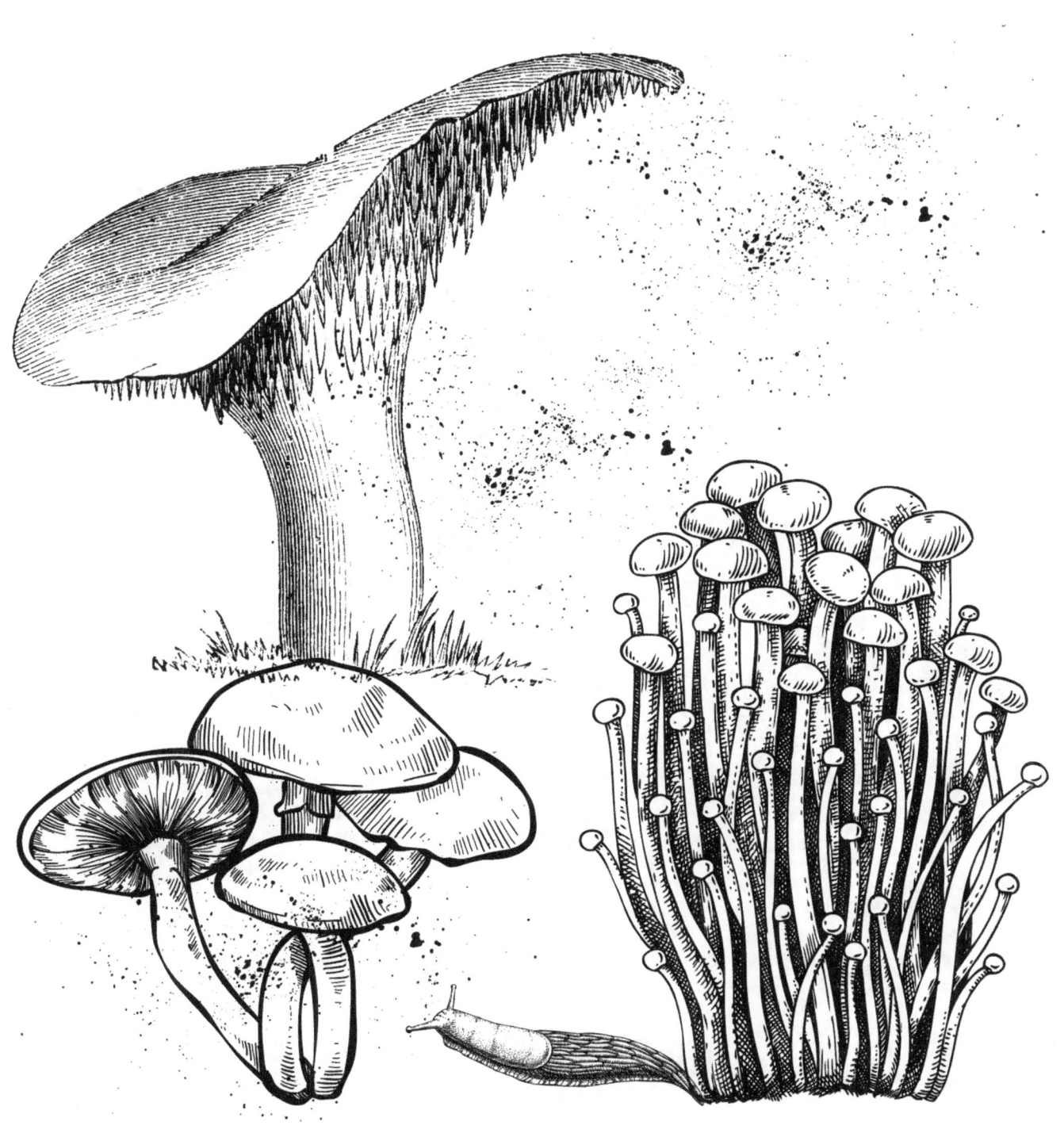

CANOPY

THE CANOPY IS THE TOP LAYER OF A FOREST,
MADE OF TALL TREE TOPS. IT HELPS THE FOREST
STAY COOL BY BLOCKING SUNLIGHT AND
REDUCING WIND AND TEMPERATURE CHANGES
BELOW.

FOREST ANIMALS

THE GIANT OTTER, NATIVE TO SOUTH
AMERICA'S AMAZON BASIN, IS CLASSIFIED
AS ENDANGERED BY THE IUCN, WITH
FEWER THAN 5,000 INDIVIDUALS LEFT IN
THE WILD.

GIANT OTTER

FOREST ANIMALS

MANY BIRDS AND INSECTS LIVE IN
FORESTS BECAUSE THEY FIND FOOD,
SHELTER, AND THE RIGHT CONDITIONS
TO STAY ALIVE.

HUMMINGBIRD

FOREST ANIMALS

SQUIRREL MONKEYS, CAPUCHINS, AND SPIDER
MONKEYS OFTEN EAT ANTS. THEY USE DIFFERENT
FORAGING METHODS, SUCH AS PICKING ANTS OFF
SURFACES OR BREAKING INTO NESTS TO REACH THE
LARVAE.

FOREST ANIMALS

THE AMAZON RIVER DOLPHIN, ALSO
CALLED THE PINK RIVER DOLPHIN OR
BOTO, IS A TYPE OF WHALE THAT LIVES
ONLY IN SOUTH AMERICA.

FOREST ANIMALS

GREEN ANACONDAS LIVE IN THE
RAINFORESTS OF SOUTH AMERICA,
ESPECIALLY IN THE AMAZON
RAINFOREST.

FOREST ANIMALS

THE GAINT AFRICAN SNAIL IS AN INVASIVE SPECIES IN FLORIDA. THEY CAN GROW TO BE UP TO 8-INCHES LONG.

FOREST ANIMALS

RACCOONS USUALLY LIVE IN TEMPERATE FORESTS BUT CAN ALSO ADAPT TO WETLANDS, CITIES, AND MOUNTAINS.

FOREST ANIMALS

FOREST ANIMALS

BROWN BEARS ARE WIDELY DISTRIBUTED AND CAN BE FOUND IN A VARIETY OF HABITATS, INCLUDING FORESTS (LIKE TEMPERATE DECIDUOUS FORESTS), MOUNTAINS, GRASSLANDS, AND EVEN DESERTS.

FOREST ANIMALS

MOOSE LIVE IN COLD FORESTS IN NORTH AMERICA AND EUROPE, ESPECIALLY NEAR LAKES AND STREAMS WHERE THEY CAN FIND PLANTS TO EAT.

FOREST ANIMALS

BOBCATS LIVE IN FORESTS, SWAMPS, AND EVEN DESERTS ACROSS NORTH AMERICA.

FOREST ANIMALS

THE GIANT PANDA LIVES IN BAMBOO
FORESTS IN THE MOUNTAINS OF
CENTRAL CHINA.

FOREST ANIMALS

WILD BOARS LIVE IN FORESTS
ACROSS EUROPE, ASIA, AND
PARTS OF NORTH AFRICA.

FOREST ANIMALS

BENGAL TIGER IS ONE OF THE MOST POWERFUL FOREST ANIMALS AND LIVES IN THE SUNDARBANS FOREST, A HUGE MANGROVE FOREST IN INDIA AND BANGLADESH.

CARBON SINK

CARBON DIOXIDE (CO2)
IS A GAS THAT COMES
FROM CARS, FACTORIES,
AND PEOPLE BREATHING.

A CARBON SINK OCCURS WHEN A
NATURAL OR ARTIFICIAL RESERVOIR
ABSORBS AND STORES MORE
CARBON DIOXIDE (CO2) FROM THE
ATMOSPHERE THAN IT RELEASES.

THIS PROCESS HELPS TO REDUCE
THE CONCENTRATION OF CO2, A
MAJOR GREENHOUSE GAS, IN THE
ATMOSPHERE, THUS MITIGATING
CLIMATE CHANGE.

CARBON SINKS

TREES AND FORESTS TAKE IN
CARBON DIOXIDE AND HELP KEEP
THE AIR CLEAN.

CARBON SINKS

CARBON SINKS

WHILE BIRDS AREN'T CARBON SINKS LIKE TREES, THEY HELP MAINTAIN HEALTHY ECOSYSTEMS—ESPECIALLY BY SPREADING SEEDS-WHICH SUPPORTS FORESTS THAT STORE CARBON.

TOCO TOUCAN

CARBON SINKS

THE INDIAN OCEAN, LIKE ALL OCEANS, ABSORBS CARBON DIOXIDE (CO_2) FROM THE ATMOSPHERE AND HELPS SLOW DOWN CLIMATE CHANGE.

RED-NECKED FALCON

CARBON SINKS

SOIL CAN STORE CARBON FROM DEAD
PLANTS AND ANIMALS, HELPING TO
KEEP IT OUT OF THE AIR.

CARBON SINKS

WORMS CAN HELP CREATE CARBON
SINKS, ESPECIALLY IN SOIL - BREAKING
DOWN DEAD PLANTS AND ORGANIC
MATTER IN THE SOIL.

DEFORESTATION

DEFORESTATION OCCURS WHEN FORESTS ARE CLEARED OR REMOVED, OFTEN PERMANENTLY, FOR OTHER LAND USES. IT CAN HAPPEN AT ANY TIME BUT IS DRIVEN BY SPECIFIC HUMAN AND NATURAL ACTIVITIES.

DEFORESTATION

DEFORESTATION CAUSES MONKEYS TO LOSE THE TREES WHERE THEY LIVE AND PLAY.

DEFORESTATION

DEFORESTATION CAUSES PARROTS TO LOSE THE TALL TREES THEY LIVE IN AND THE FRUITS THEY EAT.

DEFORESTATION

DEFORESTATION FORCES SLOTHS TO LOSE
THE TALL TREES WHERE THEY LIVE, AND
SINCE THEY MOVE SLOWLY, THEY STRUGGLE
TO FIND NEW HOMES.

DESSERTIFICATION

DESERTIFICATION OCCURS WHEN LAND THAT USED TO GROW CROPS TURNS INTO DRY, DESERT-LIKE GROUND.

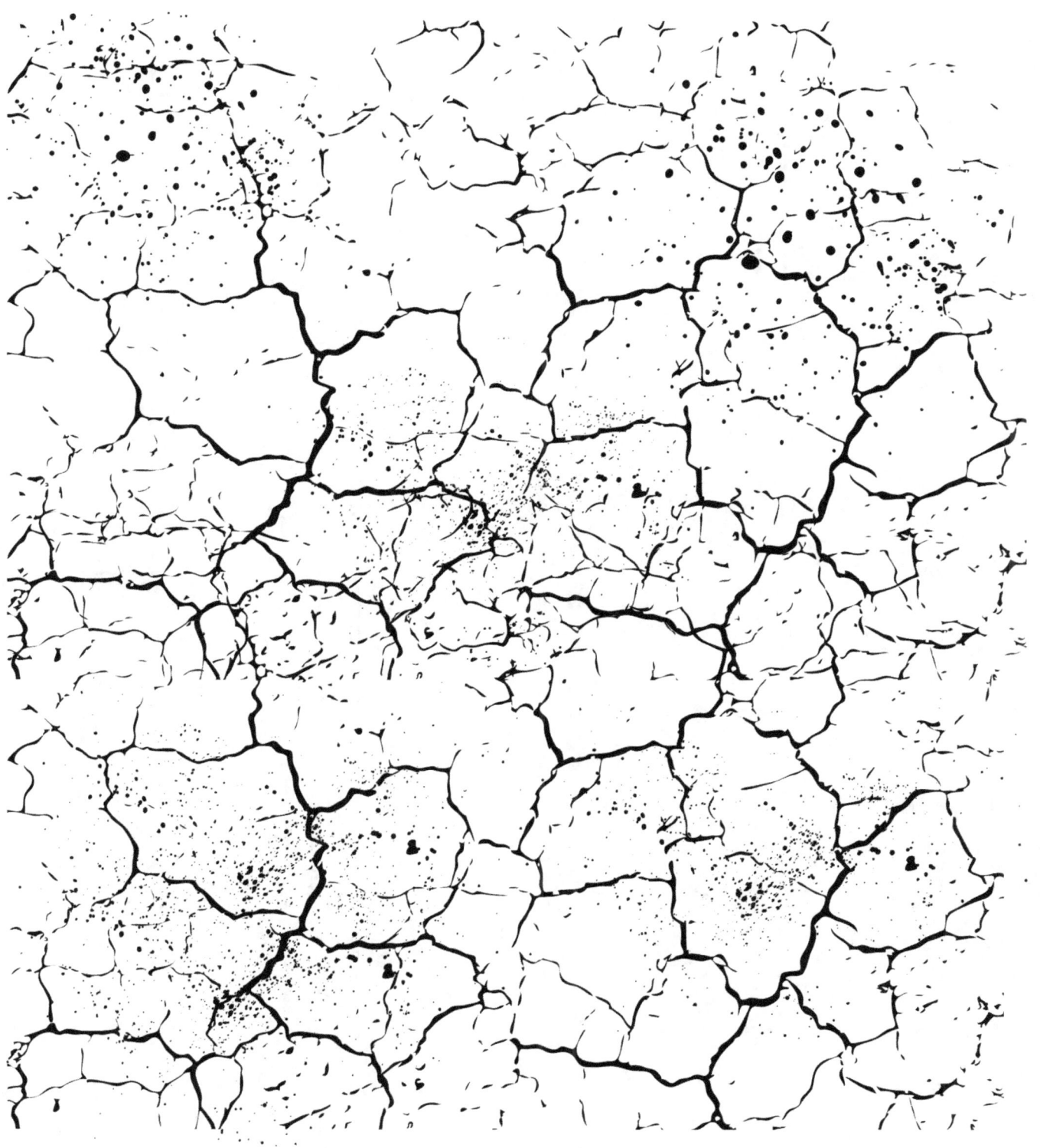

DESSERTIFICATION

DESERTIFICATION REDUCES PLANTS FOR
GRAZING AND DRIES UP WATER SOURCES,
HARMING CATTLE HEALTH AND SURVIVAL.

FOREST ECONOMICS

FOREST ECONOMICS IS THE STUDY OF HOW
FORESTS ARE MANAGED, USED, AND VALUED AS
ECONOMIC RESOURCES.

FOREST ECONOMICS

FOREST ECONOMICS LOOKS AT HOW FOREST PRODUCTS - LIKE TIMBER, RUBBER, AND NON-TIMBER GOODS - CONTRIBUTE TO ECONOMIES, JOBS, TRADE, AND SUSTAINABILITY.

FOREST ECONOMICS

FOREST ECONOMICS CONSIDERS THE COSTS AND
BENEFITS OF CONSERVATION, DEFORESTATION,
AND ECOSYSTEM SERVICES (LIKE CARBON
STORAGE OR WATER REGULATION).

"I TOOK A WALK IN THE WOODS AND CAME OUT TALLER THAN THE TREES."

-HENRY DAVID THOREAU

NOTES